Molecular Biology Techniques Applied to Assisted Reproduction

Joseph Kariuki

Bibliografische Information der Deutschen Nationalbibliothek:

Die Deutsche Nationalbibliothek verzeichnet diese Publikation in der Deutschen Nationalbibliografie; detaillierte bibliografische Daten sind im Internet über http://dnb.d-nb.de abrufbar.

ISBN: 9783346632524
Dieses Buch ist auch als E-Book erhältlich.

Das Buch bei GRIN: https://www.grin.com/document/1189883

Molecular Biology Techniques Applied to Assisted Reproduction

By

Joseph Kariuki

Contents

1. Introduction

1.1 Background to the study

According to the World Health Organization, infertility affects may couples worldwide with millions having to live with infertility issues. It is estimated that nearly 186 million people, and 48 million couples worldwide contend with the issue of infertility (WHO, 2021). It therefore presents a worrying trend for many people who wish to have children of their own. Infertility affects both women and men, with the population witnessing different prevalence's in the two genders. For instance, in North American 4.5 – 6 percent of men are affected by male infertility, while about 12% of women aged between 15-44 years will experience fertility issues. This shows therefore that there are differences in how fertility affects both males and females.

Infertility can cause serious mental torture, with the individual who is unable to get a child met with suspicion and stigma. This is actually true in developing countries, where the issue of infertility is closely associated with traditional beliefs, and stigmatization is directed on the side of the woman in the relationship. The medical advancement in developing countries is also lagging behind which makes it challenging and expensive for couples to get treatment for infertility. It makes it virtually impossible for the vulnerable in the society to get an effective solution to their problem when the health resources in the country are almost non-existent.

Since the infertility issues affects a significant portion of the population, the medical world has sought lasting solutions that could help couples who are unable to conceive to get a chance to get pregnant, and get their own children. The quest to find a solution for couples affected by infertility issues necessitated molecular biologists to find techniques that could be used to help them conceive and get pregnant. There has been advancement in the field, with history dating as back as 1978 when the first child to be sired using assisted reproductive means was born. Since then, there has been an explosion of and rapid advancement in the field of assisted reproduction, which has helped many people achieve the dream of having children.

In recent decades, there have been a number of significant advances in reproductive and microbiology that have effectively made these two fields inexorably intertwined. Today, molecular biology is vital in all stages of development, from diagnostic procedures to the selection of the most cutting-edge drugs. Genetic screening, in particular, is utilized in assisted reproduction for three key purposes: diagnosing the causes of infertility, discovering genetic disorders that may be passed down to children, and perfecting assisted reproductive technology (ART) The total fertility rate is dropping; for example, in the United States, 13% of women get fertility treatment at some time in their life. As a consequence, it is vital to support reproductive journeys of couples. To have a child, both spouses' reproductive systems must be integrated and properly coordinated; consequently, evaluating both people in the relationship is critical. A medical evaluation is essential when a couple is unable to conceive one year after indulging in unprotected intercourse. Currently, the clinical chronology of infertile couples comprises molecular and operational studies that allow for a diagnosis in 65 percent of cases; hereditary testing is undertaken in the other 35 percent of instances that go unnoticed. Recognizing that approximately 15% of genetic illnesses are associated with reproductive disorders and that relevant medical symptoms can be caused by both non-genetic and genetic causes, it is critical that a fertility problems diagnosis be made using a combination of the clients' specific medical records and instrument- and laboratory-based evaluations, including targeted genetic screening. Genetic screening (testing and counseling) to confirm a particular diagnosis may aid in the provision of more precise and focused healthcare.

The aim of this paper will be an attempt to review available molecular biology techniques that can be used to offer effective treatment solution for couples who are unable to get pregnant. Due to a wide range of options available, it is important to find evidence-based techniques that are proven to offer positive results.

1.2 Statement of the problem

Many couples and people around the world experience fertility issues, which affects their ability to conceive children. A majority of the affected people do not know where to get appropriate treatment, which further worsens their mental burden. There are available treatment solutions available to them, but many are not backed by research or science and can prove to be dangerous. This therefore necessitates the need to find appropriate molecular biology techniques applied to assisted reproduction, which can be helpful to affected couples and individuals.

2. Literature Review

In this literature review will be an analysis of studies covering on the topic of molecular biology techniques in assisted reproduction. The first part of this chapter will be an explanation of the causes of male and female infertility. This will be followed by a description of molecular biology methods that are used to assist couples who have been unable to conceive to get pregnant.

2.1 Male Infertility

Cells that are going to become germ are supposed to normally take on different destiny from other cells that are in the embryo as early as possible and particularly throughout embryogenesis. Germ cells are meant to sustain species by transmission of genetic elements to next generation. The male germ cells go through meiosis and morphogenesis in an extensive manner so that they can alter the cells which are round shaped to sperm cells that can travel in a free manner driven by flagellum and moving to seek the missing half. The additional genes and regulatory systems are needed so that the unique traits are got. If a significant number of male fertility genes are necessary in mammals, disruptive mutations in those genes are related with an elevated risk of male fertility disorders.

Chromosomal anomaly is an abnormality is in men where some males have severe male factor infertility and is detected by checking the blood chromosomes (Karyotype)so that they can have an extra x chromosome. This indicates that instead of the males having 46XY karyotype, what they have is 47XXY karyotype. This kind of illness is known as Klinefelter Syndrome and will bring about inability to achieve puberty and even if it is achieved the guy would be infertile. Men with Klinefelter Syndrome can make someone pregnant by use of in vitro fertilization (IVF) and this is done by use of Intra-Cytoplasmic Sperm Injection (ICSM) (ICSM)

The Y chromosome micro-deletions are where certain males have poor sperm count and this is because they have deletions in specific areas of their Y chromosome and this is known as DAZ gene. The karyotype of the males is normal with 46XY; nevertheless, a thorough study of Y chromosome will indicate that some of the pieces are missing. A portion of these male will have no sperm in testicular surgery and hence the only option is donation cells (Silber et al 1994). When there are other deletions in DAZ gene there is a little number of sperm that is present and hence conception can be obtained with IVF-ICFSI. The male offspring that has acquired the father's genes will eventually be infertile too for they will carry the Y chromosome.

Most of the causative divisions of male fertility have been tied to hereditary inheritance. Over than 200 hereditary disorders have been related to sexual dysfunction, varying from the most frequent clinical presentations of infertility to the strangest complicated abnormalities with clinical symptoms that stretch across fertility issues, as per our study. According to Begum (2008) infertility is generally only one of numerous clinical indications of a severe condition; nonetheless, in certain inherited diseases, infertility is the most prominent phenotypic trait. Additionally, it is important to follow these infertile individuals over time since they have been proved to have a larger morbidity rate and a low life expectancy than the overall population. In scenarios of severe oligospermia, the development of variations in the spermiogram is now the major indicator for genetic testing, especially in hormonal levels, anomalies, frequent abortions, and genetic predisposition. Recent research revealed that the percentage of genes indisputably linked to the more prevalent phenotypes of oligozoospermia or azoospermia remain restricted (50 percent); the other half are genes linked in teratozoospermia, despite the monomorphic forms of teratozoospermia being highly rare. Entire chromosomal abnormalities (structural or numerical), incomplete chromosomal aberrations (microdeletions of the Y chromosome), and monogenic diseases are all genetic disorders associated to male infertility. Sex chromosomal deficits have a larger influence on spermatogenesis, whereas genetic mutations are more connected to hypogonadism, teratospermia, or asthenozoospermia, as well as familial variants of obstructive azoospermia.

2.1.1 Whole chromosomal aberrations

First most similar genetic indicators used to detect infertility in men and are the karyotype, the evaluation of chromosomes Y chromosomal abnormalities, and the analysis of the CFTR gene. Because several abnormalities have been linked to male infertility, it is not unusual for the biologically predetermined cause of 40% of all male infertility to remain unanswered. It's also important to note that the role of de novo mutations should be extensively investigated, especially in the light of what occurs with Klinefelter syndrome and AZF deletions, which occur almost entirely. As a result, in order to optimize and customize the whole testing and therapy route of male infertility, targeted genetic variants should be conducted in the detection of specific clinical illustrations, every after suitable genetic counselling and primarily for clinical diagnosis, during clinical decision-making to classify the most effective ART approach (for instance, mostly in existence of deletions of the AZFa and AZFb zones, the possibility of sperm restoration should be considered) (for example, in the presence of deletions of the AZFa and AZFb regions, the possibility of sperm recovery using testicula). If it comes to general genetic abnormalities, the incidence of genetic variations varies in newborns (Begum,2008). When it comes to numerical abnormalities, structural chromosome inversions are more common; nevertheless, this will not apply to y chromosome, whose aberrations, chromosomal aberrations, are characterized by aneuploidies and structural chromosomal rearrangements of chromosome Y.

Klinefelter syndrome (karyotype 47, XXY) is the most common type of sex chromosomal aneuploidy seen in infertile men (karyotype 47, XXY). The existence of Y chromosome symbol that represents is the defining feature of the second most frequent gonosomal abnormality, Double Y syndrome or Jacobs syndrome. In regards to providing a lower fertility, people with congenital defects are much more likely to have had an abortion or have a kid with an abnormal karyotype. Minear et al (2015) suggests basic genetic abnormalities that may interfere with appropriate reproduction, as well as crucial morphological data, analytical procedures to identify them, and grounds for prenatal genetic testing

2.1.2 Partial chromosomal aberrations

First most similar genetic indicators used to detect infertility in men and are the karyotype, the evaluation of chromosomes Y chromosomal abnormalities, and the analysis of the CFTR gene. Because several abnormalities have been linked to male infertility, it is not unusual for the biologically predetermined cause of 40% of all infertility to remain unresolved (Begum ,2008). It's also important to note that the role of de novo mutations should be extensively investigated, especially in the light of what occurs with Klinefelter syndrome and AZF deletions, which happen almost entirely. As a result, in order to optimize and customise the whole testing and therapy route of male infertility, targeted genetic variants should be conducted in the detection of specific clinical illustrations, every after suitable genetic counselling and primarily for clinical diagnosis, throughout clinical decision-making to classify the most effective ART technique (for instance, mostly in presence of deletions of the AZFa and AZFb zones, the possibility of sperm restoration should be considered) (for example, in the presence of deletions of the AZFa and AZFb regions, the possibility of sperm recovery using testicula).

If it comes to general genetic defects, the incidence of genetic variations varies in newborns (Huang et al 2014). When it comes to numerical abnormalities, structural chromosome inversions are more common; nevertheless, this will not apply to y chromosome, whose aberrations, chromosomal rearrangements, are characterized by aneuploidies and structural chromosomal rearrangements of chromosome Y. Klinefelter syndrome (karyotype 47, XXY) is the most common type of sex chromosomal aneuploidy seen in infertile men (karyotype 47, XXY). The existence of Y chromosome symbol that represents is the defining feature of the second most frequent genosomal abnormality, Double Y syndrome or Jacobs syndrome (Devroey, 1996). In regards to providing a lower fertility, people with congenital defects are much more likely to have had a miscarriage or have a kid with an abnormal karyotype. The fundamental genetic abnormalities that may interfere with appropriate conception, as well as crucial morphological information, analytical procedures to identify them, and grounds for prenatal genetic testing

2.1.3 Single gene mutations

In single genetic variations is acknowledged that single gene defects have clinical consequences for male infertility. Although millions of genes are related with male infertility, today, only a handful of hereditary ailments are constantly explored such as cystic fibrosis. As proved by WHO (2021), the attempts to pinpoint a single causal gene are not useful given that more than 2300 genes are expressed in the testis alone and that hundreds of them govern reproductive processes and may contribute to male infertility. Huang et al (2014) suggest that almost 50 percent majority of infertility cases are ascribed to one or more hereditary issues, the genetic causes remain unclear for Furthermore, the rising widespread usage of technology, such as NGS (next-generation sequencing), for both diagnostic and research purposes enable growth of knowledge(Huang et al,2014) Starting with the clinical and laboratory evaluation, the fundamental genetic issues that can interfere with healthy reproduction are noted with the objective of refining the focused genetic test in the presence of specific clinical pictures.

2.2 Female genetic infertility

In contrast to male infertility, little is understood regarding the genetic foundations of female fertility problems. Consequently, fewer precise tests are typically advised to infertile women to determine the existence of chromosomal disorders or single-gene impairments connected to their phenotypes. Furthermore, isolated fertility issues caused by genes is unusual. Syndromic disorders lead to female infertility. Up until recently, genetic diagnostics are largely performed for patients with POI, limited to chromosomal abnormalities and FMR1 transformations. The important clinical indicators and the laboratory tests that are accessible in the pre- and postnatal periods are given in detail.

2.2.1 Chromosomal aberrations

Considering that chromosomal problems considerably effect fertility and risk for miscarriages, karyotype research is commonly advised. The most significant clinical structural conditions in infertile women are translocations, both reciprocal (transfer of two output segments from differing chromosomes) or Robertsonian (which is the centric fusion of two acrocentric chromosomes) liable for blocks of meiosis and structural alterations of the X chromosome. Patients with reciprocal translocations are at a greatly heightened threat of infertility, including hypogonadotropic hypogonadism with primary or secondary amenorrhea or oligomenorrhea. The stable inversions do not convey health concerns for their carriers because they induce neither loss nor duplication of genetic material, but they can give rise to gametes in which the genetic information is unbalanced and can thus become a cause of infertility or repeated miscarriage.

Females that have a normal karyotype create a varying fraction of eggs with chromosomal abnormalities because of mistakes that occur during meiotic nondisjunction and crossing over. The three basic classes of anomalies are Polyploidy, trisomy, and 45X. It is commonly recognized that these instances rise with the age of a female. It is conceivable to analyze gametes or embryos when receiving ART owing to PGT. The efficiency of the treatment is improved after testing for aneuploid embryos and transplanting only euploid embryos.

2.2.2 Fragile X syndrome

Fragile X syndrome is an autosomal dominant genetic disorder characterized by the existence of over 200 repetitions of the CGG triplet sequence in the FMR1 (Fragile X Mental Retardation 1) gene or by a deletion affecting the FMR2 (Fragile X Mental Retardation 2) gene. Carriers of the female FMR1 transformation (where the amount of CGG repeats falls within 55 and 200) or FMR2 microdeletion indicate early ovarian failure, diminished ovarian reserves and menstrual disorder.

Together with the lineage background, in the event of ladies with identical clinical indications, the option of a molecular analysis should be considered. The most prevalent genetic contributors to POI are X-chromosome-related disorders. In restricted circumstances, the etiology involves an alteration in an autosomal chromosome. Recognizing the genetic changes in a reasonable timeframe is of utmost significance to managing the fertility alternatives and, if required, selecting a pre - implantation genetic treatment program: the purpose is to determine the exact clinical pictures in which a specific genetic test is identified and that could direct an individually tailored diagnostic– therapy treatment strategy.

2.3 Molecular Biology Techniques in Assisted Reproduction

Technological improvements have been a crucial element in developing successful treatments that could boost the possibilities of couples conceiving a baby. The concept of assisted reproduction, has been around for a while, with advances and identification of new approaches happening. Some of the existing molecular biology techniques include: In-vitro fertilization and embryo transfer (IVF & ET) and Intracytoplasmic sperm injection (ICSI).

2.3.1 In-vitro fertilization and embryo transfer (IVF)

IVF is an abbreviation for in-vitro fertilization, and is the technique of fertilization outside of the body in a sterile situation. This method was utilized to effectively treat impotence for its first time in 1977 at Bourne Hall in Cambridge, England. Thousands of babies have been conceived all over the globe as a consequence of IVF therapy. The most prevalent causes include symmetrical tubal injury or blockage, as well as insufficient or low-quality sperm to fertilize an egg (Gregorius,1989) IVF alleviates these issues by enabling conception to occur outside of the uterus in a glass dish, thus the Latin term "in vitro," which literally means "in glass."

2.3.2 Intracytoplasmic sperm injection (ICSI)

Intraacystoplasmic sperm injection is the implantation of a single mature frozen regular sperm cells into the cytoplasmic of either a mature metaphase II egg (ICSI). Ever since debut, ICSI have transformed the treating of male infertility, with exceptional pregnancy and implant rates attained in partners who formerly had no therapeutic strategies apart from donor or adopting. Initially, ICSI was used effectively in individuals who eggs could not fertilize after first being fertilized with motile spermatozoa. And that became clear that ICSI might help individuals who do not have motile spermatozoa for traditional IVF. Eventually, scientists attempted to cure low sperm men by injecting epididymis and testes (obstructive azoospermia, OA) (non-obstructive azoospermia, NOA) (Begum,2008). This was an improvement in terms of embryo, fertilization rates, and safe childbirth10-12. Prior until 1992, the great majority of severe male factor infertility was thought to be fatal. Devroey (1996) suggests that ICSI is now a standard procedure, it is now feasible to treat the entire spectrum of male infertility, from good seminal specimens via ejaculate failure to obstructive and non-obstructive azoospermia.

3. Discussion

3.1 Molecular techniques for detecting genetic illnesses that are transmissible to offspring

Perinatal mortality is thought to be caused by inherited chromosomal or genetic disorders in 20–25 percent of cases. Because of advances in scientific understanding over the last few decades, prenatal carrier detection has become increasingly popular. The identification of spouses who are at risk of passing on a certain genetic disorder to their offspring has the potential to provide expectant parents with informed procreative options. If the fertile partner carries a defective gene in one of the inherited diseases, the child would be at risk for the same sickness.

In collaboration, the American College of Obstetricians and Gynecologists developed key principles for cultural and broad population genetic testing. Based on the age of childbearing Screening involves testing for over 2000 genetic diseases and intricate anomalies, It entails psychological impairment as well as premature heart problems.

In this scenario, genetic screening is crucial for detecting the inherited hazard, successfully recommending patients, and informing clients about hereditary concerns that may affect their judgment. In reality, prenatal transmission screening provides genetic data for a variety of disorders; as a result, all spouses will make sexual decisions easily (Begum,2008). Personalized hereditary screening is an important means for improving the outcomes for females and children in both short and long term..

Presently, a range of tests are available during the early weeks of pregnancy so that they can see any illness that can be passed on from children or if there is carrier to the affected partner. Some of these strategies may be used in isolation in various prenatal stages before the embryo is incorporated in IVF procedure. Intrusive PND is typically accomplished on recovered DNA.

The molecular diagnosis for single-gene disorders is passed either by somatic alteration examination when the parent genes are recognized or by sequence investigation when the father genes are not known. Fatherhood identification and contamination studies are always carried out in accordance with the relevant diagnostic stages. The non - invasive prenatal diagnostic (NIPD) (NIPD) of single - gene illness, which can detect fetus chromosomal abnormalities in mother plasma at an early stage of pregnancy, has received increasing attention. Around ten weeks).

NIPD has recently been used for therapeutic intervention in situations of sex-related disorders like RHD. NIPD has been studied extensively in single-gene illnesses such as - thalassemia and congenital adrenal hyperplasia. The application of NIPD in clinical settings is now possible because to the pioneering technology of NGS paired with a haplotyping approach. PGT seems to use the same diagnostic logic as conventional PND, with the additional advantage of allowing for earlier detection during the embryonic period (Wood,et al 2006) Only healthy embryos are implanted in the female, obviating the necessity for medical termination. PGT includes the use of IVF treatment, which includes (a) the collection and analysis of both partners' gametes; (b) ovum fertilization by intracytoplasmic sperm injection (ICSI) (Van Steirteghem et al,1995); (c) embryo biopsy, which allows one or more cells from the blastomere or trophectoderm to be taken 3 or 5 days after fertilization; (d) molecular analysis; and (e) embryo transfer. PGT techniques are meant to start with a little quantity of biological material, generally 1 to 10 cells from a cleavage or blastocyst stage embryo. There is dispute on whether embryonic biopsies are harmful, and phenotypic plasticity arises in cleavage stage and blastocyst embryos.

Preimplantation Genetic Diagnosis International Society (PGDIS) data from 2018 demonstrated minimal variation in mostly unfavorable impacts across embryo stages offered by qualified practitioners who conducted the biopsies. PGT use whole genome amplification (WGA) to acquire a sufficient quantity of DNA Sequence for one or more future molecular methods. Different kinds of WGA may be employed effectively and contribute to the overall purpose. The most commonly used PGT technique is still "multiplex polymerase chain reaction" (PCR) (PCR) and capillary electrophoresis examination for the direct identification of the disease's causative gene and the study of at least one or two relevant morphological traits, the most common of which are "short tandem repeats" (STR), microsatellites characterized by short tandem nucleotide repetitions, or the analysis of at least three polymorphic markers.

Nonetheless, while PGT customized to a disease is a difficult and costly technology that takes time to evaluate families, other labs employ genomic sequence techniques to assess gene markers throughout the genome. Alan Handyside looked into "karyomapping," which focuses on a single gene polymorphism (SNP) matrix capable of detecting a person's genotype by analyzing hundreds of SNPs scattered across the genome. (Aslam, 2021). The "karyomapping" assessment includes a "linkage" assessment: the prevalence of the mutagenesis provider can be determined by comparing the SNPs associated with the correlational mutated gene of the ailment to be investigated, which are prevalent in the affected individual and thus in the parents' chromosomes, to the SNPs present in the fertilized egg cells. In general, the concentration of SNPs provides more clarity in the condition of chromosomal crossings surrounding the mutant gene. Finally, "karyomapping" may be employed in conjunction with other techniques. Further single-gene mutations or those needing HLA concordance should be rigorously tested using traditional techniques. "Karyomapping",

However, it is ineffective if the paternal gene is unrelated to the genetic alteration; for empirical testing of mitotic aberrations (mosaicisms), "karyomapping" does not directly evaluate the modification and so cannot detect de novo mutations. The most current approach is to uncover single mutations or clusters of common variations using targeted SNP research, which is then matched with statistical haplotype evaluation. or the number of chromosomes In combination with cytogenetic testing, a genomic sequence technique based on NGS has recently been investigated for the diagnosis of family abnormalities in embryonic biopsies. Based on a large panel of disease genes (approximately 5000 genes), the method allows for (a) the direct identification of hereditary genetic variations and the indirect identification of heterozygous SNPs by genetic markers (PGT-M); (b) a chromosomal homologous recombination (PGT-SR) study; and (c) testing for aneuploidies in a single operation. Taking into mind the limits of a single NGS technique, the inability to identify haploidies, polyploidies, and mosaicisms has been ascribed to it. Furthermore, investigating consanguineous relatives is not advised.

Finally, further limitations such as the detection limit or the amount of translocation permitted by the approach may be circumvented by applying haplotyping in the context of the indicator instance.

4. Advancement in Molecular Biology in Assisted Reproduction

PGT is a complex laboratory procedure that necessitates interprofessional research between ART practitioners and research lab geneticists who have expertise analyzing small quantities of cell samples. The differentiation among the two primary PGT techniques, PGS and PGD, is already disappearing from a technological viewpoint, because MPS performs genome-wide SNV and CNV variant hereditary lab screening concurrently. According to Zegers-Hochschild et al. (2017) this is stressed by the International Committee for Monitoring Assisted Reproductive Technology's recommendation to update the language in the "International nomenclature on infertility and reproductive therapy" for several laboratory tests. As a consequence, PGT would serve as the main procedure term to which PGT monogenic disease evaluation (PGT-M) and chromosomal aneuploidy analysis (PGT-A), as well as all other genetic screening methodologies, might be linked. Traeger-Synodinos, (2017) argued that PGT-M is intended to detect a distinct 'Mendelian' herediatry abnormality in the embryo where the intended guardians of the child will be highly predisposed. PGT-M is superior to normal PND in that it does not necessitate ETP assessment in a complication-free prolonged pregnancy.

PGT- A is a method for identifying chromosomal aneuploidy in order to choose or selectively analyze embryos that do not show any signs of having disease-related chromosomal defects. These screened and selected embryos are thought to have the highest likelihood of growing into a successful live delivery. According to Geraedts and Sermon (2016) PGT-A is primarily used to increase the success rate of IVF. Dahdouh et al. (2015) also show that PGT utilizing 'full chromosomal screening' procedures enhanced medical and maintained implantation rates, especially in females with standard ovarian reserves. Harper et al., (2017) suggests that because the research in the aforementioned study from Dahoud et al. only applied to patients with a good prognosis, further evidence is required to support the efficacy of PGT-A in promoting positive clinical outcomes in new patient groups and at other phases of embryo biopsy. Geraedts and Sermon, (2016) advice that it is vital to highlight that because 'IVF success' is perceived and described differently by several authors, comparing the findings of several research from evidence-founded meta-analyses is nearly difficult. Improved patient categorization before PGT-A would assist in determining whether the testing approach is best for different patient groups.

As a result, because the findings of standardized RCTs have yet to be published, there is insufficient data to ascertain the favorable benefits of PGT-A. (Harper et al., 2017). Because PGT-A is costly, most European healthcare systems utilize 'positive' RCT findings as a criterion for PGT-A funding. With the development of better vitrification technologies (Vajta et al., 2015) and the completion of a particular RCT, frozen embryo transfer has gradually surpassed fresh embryo transfer (Coates et al., 2017). This technique allows for additional time for high-quality PGT while also allowing for the gathering of increased 'diagnostic cases' for concurrent evaluation, that saves money.

(Geraedts and Sermon, 2016) also add that greater usage of PGT backed by evidence and research findings is linked to advancements in embryo vitrification and is contingent on top quality expertise in the micromanipulation of embryos. Thornhill et al. (2015) show that full-genome haplotyping methods, such as Karyomapping. together with MPS-based whole-genome 'deep sequencing (Esteki et al., 2015) allow for simultaneous CNV analysis, SNV analysis, and haplotyping and aneuploidy evaluation.

Haplotyping allows for effective PGT-M for nearly any inherited illness-linked variant (even without the need for an earlier effort of analyzing the relevant variants in a particular family). (Zheng et al., 2015; Sermon et al., 2016) describe that genome-wide low-coverage MPS enables

PGT-A, and includes the diagnosis of small chromosomal aneuploidies as small as 1.8 Mbp. Many PGT-M testing techniques historically focused on evaluating a specific confined portion of the genome; however, new technological methodologies may allow for a more comprehensive analysis of an embryo's genetic diversity. Therefore, not only significant genetic variances are revealed, but also the variation in genomes that are not related to the pair's initial referral and request.

5. Ethical Considerations

Assisted reproductions technologies discussed draw attention to specific ethical difficulties that need to be considered (Van den Veyver, 2016), and expert recommendations must be made as a result. For instance, PGT, like all other types of clinical DNA testing, should be performed under stringent quality protocols and evaluation. Frequent involvement in quality assessment must be perceived as necessary, and in certain countries, it is even needed for genetic testing. PGT-specific techniques for aneuploidy and monogenic testing are widely available in European countries and in other global regions. Harper et al. (2010) suggest that the most efficient technique of quality assurance remains laboratory accreditation in line with ISO15189 and its popularity is expanding throughout Europe and other continents.

Guidelines that are unique and intended for a particular like those used for MPS diagnosis should be applicable, and considered as mandatory for PGT. These guidelines should also be applied at all times to ensure that there is no laxity that arises in gene testing in any laboratory (Matthijs et al., 2016). Richards et al. (2015) also adds that, The Standards and recommendations for sequence variant interpretation": a unified recommendation from the Association for Molecular Pathology and the American College of Medical Genetics and Genomics should also be adhered to. In addition, in PGT, Human Genome Variation Society (HGVS) guidelines on sequence variant terminologies and correct assigning of 'disease association' in identified variants (therefore leaving the prefix 'pathogenic' only to an evident set of diagnostic scenarios highlighting that a given variant 'leads to illness only in a particular setting) (HGVS, 2017), as well as relevant diagnostic test reporting recommendations, should be followed (Claustres et al., 2014).

Embryo mosaicism, which prohibits categorical (or 'conclusive') results from being presented, is another impediment to practical PGT-A implementation. Although the trophectoderm genetic test allows for the collection of 'summative' template DNA from numerous embryonic cells instead than a few within a blastomere ('Day 3') biopsy, it is still a normal biopsy method with inherent shortcomings in terms of sample of the organ and tissues under investigation.

Kamps et al., 2017 suggests that these constraints are the same to those being addressed in the area of cancer research, where precise diagnosis and reproducible study discoveries rely greatly on the biopsy operator's expertise and accurate positioning of the sample inside the tissue of interest. Also, (Fragouli and Wells, 2011) in their study concluded that because mosaicism has been found to be prevalent at the blastocyst stage, an embryo biopsy sample (in any phase of development) may not be generalizable of the complete embryo.

6. Conclusion

In summary according to Scott and Galliano, (2016), whereas MPS enhances evaluation sensitivity and hence mosaicism identification in a single embryo biopsy, it is not effective in overcoming the restriction of just analyzing the cells in the sample. Greco et al., (2015) however show there can be successful pregnancies after the replacement of chosen mosaic aneuploidy blastocysts. This discovery, which has the potential to shift the present PGT-A paradigm, demands more multicentric inquiry into the initial observation, as well as optimum embryo prioritizing based on qualitative and quantitative features of embryo chromosomal mosaicism.

The existence of mosaicism, as well as the kind of aneuploidy (chromosomal, monosomy, or trisomy), determines the success rate of having a healthy live delivery. The ongoing problems in this field is to determine 'cut-off' levels that will allow embryos to be classified as transferrable as first priority (which indicates that is no detection of mosaicism), transferrable as second priority (which indicates there is moderate mosaicism with exclusion of chromosomes potentially related to viable pathological conditions), or transferrable as third priority (moderate mosaicism with exclusion of chromosomes potentially related to viable pathological conditions) (PGDIS). As a result, the PGT-A is primarily a "ranking tool" for instance it can be seen as a quantitative tool that enhances decision making.

References

Aslam, F., Yasmin, S., Reza, S., Shahid, A., Mahmud, M., & Mahmud, A. (2020). Association of Pituitary Hormones and Anthropometric Parameters in Obese Infertile Women. *International Journal of Pathology*, *18*(4), 136-141.

Aslam, F., Yasmin, S., Reza, S., Shahid, A., Mahmud, M., & Mahmud, A. (2020). Association of Pituitary Hormones and Anthropometric Parameters in Obese Infertile Women. *International Journal of Pathology*, *18*(4), 136-141.

Begum, M. R. (2008). Assisted reproductive technology: Techniques and limitations. *Journal of Bangladesh College of Physicians and Surgeons*, *26*(3), 135-141.

Claustres, M., Kožich, V., Dequeker, E., Fowler, B., Hehir-Kwa, J. Y., Miller, K., ... & Barton, D. E. (2014). Recommendations for reporting results of diagnostic genetic testing (biochemical, cytogenetic and molecular genetic). *European journal of human genetics*, *22*(2), 160-170.

Coates, A., Kung, A., Mounts, E., Hesla, J., Bankowski, B., Barbieri, E., ... & Munné, S. (2017). Optimal euploid embryo transfer strategy, fresh versus frozen, after preimplantation genetic screening with next generation sequencing: a randomized controlled trial. *Fertility and sterility*, *107*(3), 723-730.

Dahdouh, E. M., Balayla, J., & García-Velasco, J. A. (2015). Comprehensive chromosome screening improves embryo selection: a meta-analysis. *Fertility and sterility*, *104*(6), 1503-1512.

Devroey, P., Nagy, P., Toumaye, H., Liu, J., Silber, S., & Van Steirteghem, A. (1996). Outcome of intracytoplasmic sperm injection with testicular spermatozoa in obstructive and non-obstructive azoospermia. *Human Reproduction, 11*(5), 1015-1018.

Esteki, M. Z., Dimitriadou, E., Mateiu, L., Melotte, C., Van der Aa, N., Kumar, P., ... & Voet, T. (2015). Concurrent whole-genome haplotyping and copy-number profiling of single cells. *The American Journal of Human Genetics, 96*(6), 894-912.

Fragouli, E., & Wells, D. (2015, November). Mitochondrial DNA assessment to determine oocyte and embryo viability. In *Seminars in reproductive medicine* (Vol. 33, No. 06, pp. 401-409). Thieme Medical Publishers.

Geraedts, J., & Sermon, K. (2016). Preimplantation genetic screening 2.0: the theory. *Mhr: Basic science of reproductive medicine, 22*(8), 839-844.

Greco, E., Minasi, M. G., & Fiorentino, F. (2015). Healthy babies after intrauterine transfer of mosaic aneuploid blastocysts. *The New England journal of medicine, 373*(21), 2089-2090.

Gregorios, A. (1989). General for high coptic studies, culture and scientific research. Christianity views on IVF and ET in "Treatment of infertility" and test tube babies. Kamal R, ed. *Akhbar El Youm, 82*, 131.

Harper, J., Jackson, E., Sermon, K., Aitken, R. J., Harbottle, S., Mocanu, E., ... & Lundin, K. (2017). Adjuncts in the IVF laboratory: where is the evidence for 'add-on'interventions?. *Human Reproduction, 32*(3), 485-491.

HGVS. Human Genome Variation Society. (2017). Available from: http://varnomen.hgvs.org.

Huang, J. Y. J., & Rosenwaks, Z. (2014). Assisted reproductive techniques. In *Human Fertility* (pp. 171-231). Humana Press, New York, NY.

Infertility, W. H. O. (1991). A tabulation of available data on prevalence of primary and secondary infertility. *Programme on material and child health and family planning division of family health. Geneva: World Health Organization.*

Kamps, R., Brandão, R. D., Bosch, B. J., Paulussen, A. D., Xanthoulea, S., Blok, M. J., & Romano, A. (2017). Next-generation sequencing in oncology: genetic diagnosis, risk prediction and cancer classification. *International journal of molecular sciences, 18*(2), 308.

Minear, M. A., Alessi, S., Allyse, M., Michie, M., & Chandrasekharan, S. (2015). Noninvasive prenatal genetic testing: current and emerging ethical, legal, and social issues. *Annual review of genomics and human genetics, 16,* 369-398.

Scott, R. T., & Galliano, D. (2016). The challenge of embryonic mosaicism in preimplantation genetic screening. *Fertility and sterility, 105*(5), 1150-1152.

Sermon, K., Capalbo, A., Cohen, J., Coonen, E., De Rycke, M., De Vos, A., ... & Geraedts, J. (2016). The why, the how and the when of PGS 2.0: current practices and expert opinions of fertility specialists, molecular biologists, and embryologists. *MHR: Basic science of reproductive medicine, 22*(8), 845-857.

Silber, S. J., Van Steirteghem, A., Nagy, Z., Liu, J., Tournaye, H., & Devroey, P. (1996). Normal pregnancies resulting from testicular sperm extraction and intracytoplasmic sperm injection for azoospermia due to maturation arrest. *Fertility and sterility, 66*(1), 110-117.

Traeger-Synodinos, J. (2017). Pre-implantation genetic diagnosis. *Best Practice & Research Clinical Obstetrics & Gynaecology, 39,* 74-88.

Vajta, G., Rienzi, L., & Ubaldi, F. M. (2015). Open versus closed systems for vitrification of human oocytes and embryos. *Reproductive biomedicine online, 30*(4), 325-333.

Van den Veyver, I. B. (2016). Recent advances in prenatal genetic screening and testing. *F1000Research, 5*.

Van Steirteghem, A., Tournaye, H., Van Der Elst, J., Verheyen, G., Liebaers, I., & Devroey, P. (1995). Intracytoplasmic sperm injection: Three years after the birth of the first ICSI child. *Hum Reprod, 10*(October), 2527-2528.

Wood, C., Trounson, A., Leeton, J. F., Renou, P. M., Walters, W. A., Buttery, B. W., ... & Victor, Y. H. (1983). Clinical features of eight pregnancies resulting from· in vitro fertilization and embryo transfer. *Fertility and Sterility, 39*(4), 98-105.

Zegers-Hochschild, F., Adamson, G. D., Dyer, S., Racowsky, C., De Mouzon, J., Sokol, R., ... & Van Der Poel, S. (2017). The international glossary on infertility and fertility care, 2017. *Human reproduction, 32*(9), 1786-1801.

Zheng, H., Jin, H., Liu, L., Liu, J., & Wang, W. H. (2015). Application of next-generation sequencing for 24-chromosome aneuploidy screening of human preimplantation embryos. *Molecular cytogenetics, 8*(1), 1-9.

5

Unidades .

The aim of this paper will be an attempt to review available molecular biology techniques that can be used to offer effective treatment solution for couples who are unable to get pregnant. Due to a wide range of options available, it is important to find evidence-based techniques that are proven to offer positive results. Many couples and people around the world experience fertility issues, which affects their ability to conceive children. (...)

Dokument Nr. V1189883
https://www.grin.com
ISBN 9783346632524